Contents

A lifeless lake	4
How rain becomes acid	6
Air pollution and the wind	12
Acid rain in the environment	14
Solving the problems	26
International action	36
What can we learn?	40
Glossary	44
Finding out more	46
Index	48

Words printed in **bold** in the main text are explained in the glossary on page 44.

A lifeless lake

Let's begin by visiting Sweden, a country in **Scandinavia**. Here, you can visit many lakes where the water is clear and beautiful. But you will not see any fish or plants in some of these lakes because the water is too **acid**. Sadly, many lakes in Sweden and other countries have been poisoned by acid rain.

▼ *This lake in Canada is clear and beautiful but it has no life in it. The water is too acid.*

▲ *Rainwater looks clean and pure, but in some places it has become acidic.*

Acid gases are produced when **fossil fuels** like coal and oil are burned in factories and in our homes. The acid gases from the smoke and fumes are often blown to other countries by the wind, and may fall as acid rain.

Acid rain can affect our health and damage our buildings, our lakes and our trees. Let's learn more about acid rain and see what can be done to stop it.

How rain becomes acid

Have you heard of acid rain? Many people are talking about it because they are worried about the effects it is having on the **environment**.

All rainwater is slightly acid. It can slowly dissolve or wear down rocks into strange shapes. These rocks are in Australia. ▶

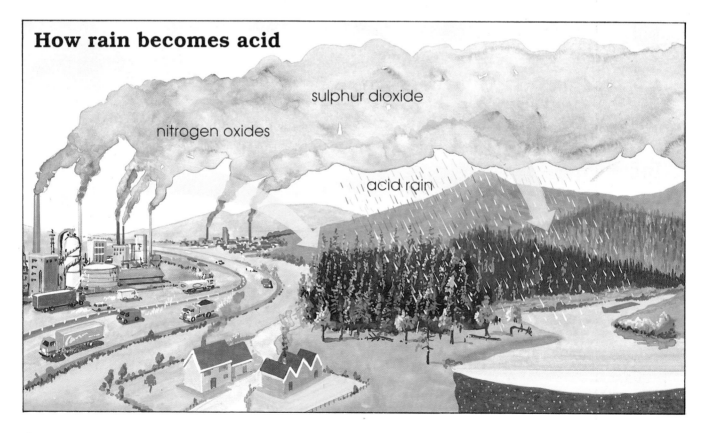

How rain becomes acid

nitrogen oxides

sulphur dioxide

acid rain

How acid is acid rain?

Did you know that lemon juice, vinegar and cola are all acid? But they are not as acid as the acid in car batteries. This is so strong it can burn holes in clothes. The opposites of acids are **alkalis**. Strong alkalis like ammonia and bleach can be dangerous too, but weak ones like toothpaste and baking powder are not. Acid rain can sometimes be as acid as lemon juice.

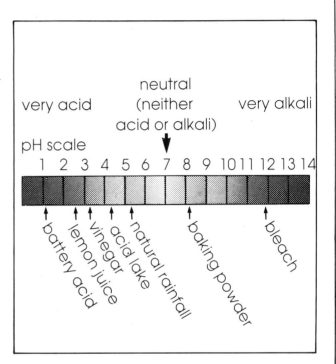

Measuring acidity

The strengths of acids and alkalis are measured on a special scale called the **pH scale**. A low pH number means something is more acid. Something that is between acid and alkali is said to be neutral. Acidity can be tested using litmus paper.

Acids turn litmus paper red, and alkalis turn it blue. With a special paper called universal indicator, you can test levels of acidity.

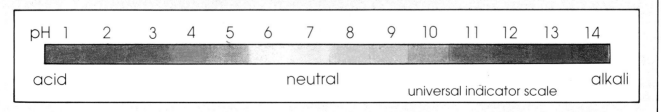

When we put harmful gases into the air it is called air **pollution**. Most air pollution results when we burn fossil fuels, such as coal and oil.

▲ *When volcanoes erupt, the gases they produce can make rain more acid. This is a volcano in Alaska, USA.*

This power station in Maryland, USA, burns fossil fuels to make electricity. ▶

How to make 'acid rain'
Ask an adult to help you with this and other experiments in this book. Pour some distilled water into a jar. Now dip in a piece of universal indicator paper. What colour is it? Very carefully light four matches and put them in the jar. They will go out and give off smoke. Take the matches out, put a lid on the jar quickly and shake it. Test again. What colour is the paper now? It should be orangey red, because the smoke has made the water acidy. Read the pH value from the scale on page 7.

When we burn fuels, chemicals called sulphur and nitrogen are released. They mix with the air to make chemicals called **sulphur dioxide** and **nitrogen oxides**, which can damage the environment.

Where do these chemicals come from? Most sulphur comes from power stations which make electricity. Most nitrogen oxides come from car exhausts. Other chemicals are released, too, but sulphur dioxide and nitrogen oxides are the main causes of acid rain.

The chemicals and moisture in air react together to make weak acids. These may be carried long distances by the wind before falling as acid rain or acid snow.

▼ *When acid snow melts, the acid water can harm plants and animals.*

▲ *Where trees are cut down and taken away, the soil can become more acid.*

Air pollution is not the only reason for the environment becoming more acid. In nature, plants rot after they die and this returns **nutrients** to the soil. But when plants and trees are cut down and taken away the soil loses these nutrients and becomes more acid.

Air pollution and the wind

▲ *Smoke from chimneys can be carried long distances by the wind.*

In 1881, a scientist found signs of pollution in an area of Norway where there was no industry. He thought it must have been carried from Britain. But it was nearly a hundred years later that scientists proved that air pollution can be carried very long distances.

A gentle wind of 16 kilometres per hour can carry pollution over 1,600 kilometres in 5 days. From Britain, this would take pollution to Sweden and beyond. The acid pollution in Sweden comes from other European countries as well as Britain. In North America, winds blow acid pollution from north-eastern USA to eastern Canada.

Acid rain in North America

Legend:
- areas slightly affected
- areas moderately affected
- areas badly affected
- • sulphur produced

▲ *This map shows the spread of acid rain in North America. The red dots show where much sulphur is produced.*

Acid rain in Scandinavia

Where sulphur is produced
- 50,000 tonnes/year
- 50,000 – 200,000 tonnes/year
- over 200,000 tonnes/year

Main direction of wind →

Areas most affected by acid rain

▶ *These three maps show how sulphur pollution produced in Western Europe affects northern countries such as Sweden and Norway. How is the pollution reaching them?*

Acid rain in the environment

Some lakes and forests in Scandinavia are badly affected by acid rain. But where soils are naturally alkaline, the lakes and forests are less affected.

For example, chalk is an alkaline rock, and the chalky soil which forms on top of this rock is also alkaline. When acid rain falls on alkaline soil, the acid is made much weaker and so there is little damage to the environment.

▼ *Many sorts of plants grow on chalky soil.*

◀ *Soils on granite rocks can be badly affected by acid rain.*

Soils and acid rain

Collect samples of different types of soil. Mix water with a few drops of vinegar. Pour some over your first soil sample using filter paper. Then test the acidity of the water with universal indicator paper.

Repeat with other samples. Do you get different results? Why is that?

Testing soil samples

When acid rain falls on the ground, small amounts of poisonous **minerals** – which are found in soil naturally – can be released. They can be taken up by plants and then passed on to animals that eat the plants. They are also washed into rivers and lakes where they can kill fish.

Millions of people enjoy fishing in their spare time. Many people earn their living through fishing. So it is important that lakes and rivers do not become too acid for fish to survive.

Lake damage

'We have a summer house by a lake. We cannot eat fish from the lake any more because they contain poisons.'

Ulf Pederson, Norway

Acid water supplies

'We live in the Swedish countryside and we always used to get water from a well. We first noticed something wrong when the fish we put in the well kept dying. When we measured the pH level, we found it was 4.9.'

Astrid Fjortoft, Sweden

A healthy lake has a pH of about 6.5.

This lake in Sweden is too acid for some kinds of fish. What pH is it? ▼

A dragonfly rests on a water lily in a healthy lake. ▶

Some facts about acid lakes

- Every year, some rain falls in Norway that is as acid as lemon juice.
- The waters of 550,000 lakes in eastern Canada are acid, or are becoming acid.
- Fish have disappeared from more than 200 acid lakes in the Adirondack Mountains, USA.

Some of the water in lakes comes directly from rainfall. But most of it passes through the ground before going into the lake.

The Osprey

The osprey is a large bird that flies low over lakes and picks out fish with its claws. But ospreys are now becoming less common. When fish in the lakes are killed by acid rain, there is no food for ospreys.

Acid levels in lakes and rivers are highest when the snow melts in the spring. This is the time when young fish hatch from their eggs. If the water is too acid the young fish will die. Other water creatures may die too.

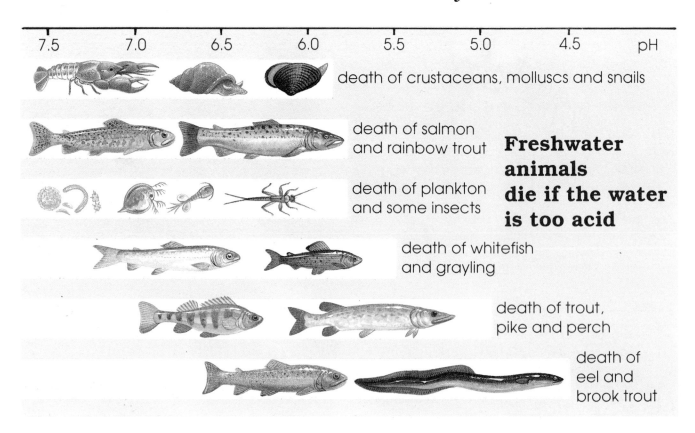

Freshwater animals die if the water is too acid

It was once thought that seas were not affected by acid rain. Then scientists found that acid rain was damaging fish along the Atlantic coast of the USA. The sea along coastlines is important because this is where fish **breed**.

▼ *New York, on the Atlantic coast of the USA, has suffered from acid rain damage.*

Coniferous trees are badly affected by acid rain. The Swedish forestry industry, which sells a lot of **timber** to other countries, is being harmed by acid rain.

▼ *Acid rain damages coniferous forests.*

The many cones on this tree are a sign that the tree is dying. It is making seeds for new trees before it dies from the effects of acid rain. ▼

Here is an experiment to show how 'acid rain' affects seed growth. ▶

Seed growth and acid rain

Sow one type of seed in 6 cups (one seed per cup) and label them. Then water 2 cups with tap or distilled water, and measure their pH. Water the next 2 cups with water and vinegar (pH 4.0), and the next 2 with water and chalk dust (pH about 8.0). Repeat the experiment using different seeds.

Compare the growth of the seedlings over several weeks.

Acid rain weakens trees, which may then be blown over or attacked by insects.

◄ *This tree branch has lost many of its needles as a result of acid rain.*

How acid rain affects coniferous trees

Fewer leaves: Healthy coniferous trees, such as fir and pine, do not drop their needle-like leaves every year. But damaged trees soon lose their needles and look bare.

Yellow spots: The sulphur dioxide in air makes the needles of conifer trees develop yellow spots.

Many cones: A damaged conifer tree produces many cones when it is dying.

Die-back: The top of a tree is more exposed to acid rain, and loses more needles than the rest of the tree.

▲ *Canadian maple trees are being damaged by acid rain.*

Forest damage

'My family has lived on this farm for over 100 years. Many of the old trees are now damaged by acid rain and are dying from insect attacks. A couple of years ago a strong wind blew some of them down. The younger trees will never reach this age.'

Ingemar Zachrisson, Sweden

If you look at old buildings, you might notice that the stone is dissolving. This happens naturally, but acid rain can speed up the process.

▲ Acid rain has damaged this English church.

The Statue of Liberty in New York had to be restored because of acid rain damage. ▼

▲ *Here in Germany, traffic is being diverted to avoid smog – a mixture of pollution and fog.*

In some places, even breathing can be dangerous for people with health problems like asthma. In Germany, pollution in the air is sometimes so bad that people are advised not to go out.

Solving the problems

More and more people are becoming worried about acid rain. Because of this, action is being taken to tackle the problem. For example, some countries now have strict pollution controls.

Join the club!

In 1984, ten countries in Europe formed a '30 per cent club'. They promised to cut down the amount of sulphur dioxide they produced by at least 30 per cent (one-third) by the year 1993. Although Britain did not join the club, it was agreed that sulphur dioxide would be cut by nearly two-thirds by 2003.

◀ *A layer of pollution drifts over a valley in Germany. The Germans quickly took action when they found that pollution was harming their forests.*

▲ *A Swedish lake is treated with lime.*

One way of fighting the effects of acid rain is by dropping lime into lakes. This is called 'liming'. Lime is an alkali made by crushing limestone rock. When it is put into lakes, the acidity of the water is reduced. Sometimes lime is spread over ice-covered lakes. When the ice melts, the lime falls into the water.

If the pH level can be raised to pH 6.5, plants and animals can return to the lake. But liming has to be repeated every 2 to 5 years and it is expensive.

◀ *These dying trees might be helped if lime were put on the soil, but it is difficult to spread lime in thick forests.*

Sometimes, lime is spread on the ground and this helps the soil and plants. But liming will not solve the problem of acid rain. We must cut down on the amount of sulphur dioxide and nitrogen oxides that are produced.

Power stations provide the electricity we need for making industrial products, heating, lighting, cooking and transport. But power stations also cause some of the acid rain. If we used electricity more carefully, we would not have to make so much, so there would be less pollution.

▼ *Power stations produce electricity, but they also produce the gases that cause acid rain.*

Much of the world's energy comes from coal, oil and gas. But electricity can also be made in ways that do not cause acid rain.

◀ *This power station in France makes electricity from the heat of the sun.*

These windmills in the USA produce electricity. ▼

▲ *This is a hydro-electric power station. It makes electricity from falling water.*

Another source of energy that does not produce sulphur dioxide pollution is nuclear power. But people worry about the dangerous **nuclear waste** that is produced.

▲ *Some power stations now have special cleaning devices to remove most of their sulphur dioxide pollution.*

Another way to reduce pollution is to take the sulphur out of coal before it is burnt. Or sulphur can be taken out of the fumes by fitting cleaning devices called **scrubbers** to power station chimneys.

When a tree is cut down for timber, it is best to take only the trunk. The roots and branches can be left to rot. This helps stop acidity building up in the soil.

▼ *This forester is cutting down a tree for timber.*

Can you think of ways in which we can all help cut down air pollution? Write them down and then look at the list on page 39.

◄ *If more people used buses and trains instead of cars, there would be less air pollution. Why is this?*

Cars are a major cause of air pollution. Using lead-free petrol is a good idea, but harmful gases are still given off from car exhausts. Most of these gases can be reduced by fitting the exhaust with a special filter. It is called a **catalytic converter**.

▲ *Using lead-free petrol is one way cars can make less air pollution.*

◄ *The exhaust fumes from vehicles contain nitrogen oxides, sulphur dioxide and other harmful gases.*

In the USA and Japan, laws have been passed to reduce pollution from cars. Similar laws will soon be passed in European countries.

International action

It costs a lot of money to clean up and prevent acid rain. But money must be spent in order to stop even more damage to the environment. Some countries have begun to take action, and the amount of sulphur dioxide produced in Europe has fallen.

Some lakes in Scandinavia are recovering a little, but air pollution needs to be cut still more.

Protesters climb a factory chimney to draw attention to the pollution it produces. ▼

The European parliament often meets to talk about environmental matters. ▶

Saving a Swedish lake

'Our lake has been saved by schoolchildren from the city. Many sorts of wildlife were disappearing because the lake was becoming acidic, and the children wanted to save it. They helped us spread lime on the ice. The next summer they measured the acidity of the water. It had fallen from pH 5.7 to pH 6.8.'

Bjorn Hansen

As well as cutting sulphur dioxide, we must also cut the amount of nitrogen oxides that we produce. About half the nitrogen oxides in air come from car exhausts.

The USA has strict laws controlling pollution from cars. Other countries have made slower progress, but all new cars in the **European Community** countries will have to be fitted with a catalytic converter by 1993.

▼ *Los Angeles in the USA is having to reduce air pollution from cars. Can you see the smog in this picture?*

Reducing pollution

Here are several ways in which we can help to cut down air pollution:

Put on extra clothes when it is cold, instead of turning up the heating. This will mean that less fuel is burnt.

Turn off the lights when they are not needed, so that power stations do not have to produce so much electricity.

Ask the drivers in your family to drive cars more slowly so that the engine produces less pollution. Drivers could also turn the engine off when a car is in a long queue or waiting at traffic lights.

Use buses and trains instead of cars – they can carry far more people in one trip. This cuts down the amount of pollution. Better still, walk or cycle whenever you can.

Can you think of any other ways?

What can we learn?

Acid rain is a big problem that is damaging our environment. But not everyone agrees about all the causes and effects of acid rain.

People often disagree about other issues affecting the environment too. It is up to us to make up our own minds, but first we must learn about the issues and listen to different points of view.

▼ *We need to learn about our environment.*

This sign in Tokyo, Japan, shows levels of air pollution.

We can all do something to help the environment. Think what you could do about this, and ask your teacher for some more ideas.

▼ *These healthy trees and lake are signs of a clean environment. Let's try to keep it healthy!*

Action at home

You can test to see if the rain where you live is acid.

Cut the top off a plastic drinks bottle, and put a plastic bag inside the bottom half as a lining.

Place the bottle outside, away from buildings and trees, and support it by tying it to a stick pushed into the ground.

When it has rained, lift out the plastic bag, and test the water with universal indicator paper. Use the colour scale on page 7 to see how acid your rainwater is. Take the sample in a screwtop jar for testing at school if necessary.

Repeat this test several times, but each time use a fresh plastic bag.

Natural rainwater has a pH of 5.6. If the pH levels of your samples are more acid than this, you have acid rain.

Each time you do the test, you should keep a note of what direction the wind comes from when it rains. If the water is more acid when the wind is from a certain direction, there may be a reason for this that you can work out. For example, has the wind come from an area with power stations or much traffic?

If the rain is acid where you live, what do you think could be done to make it less acid?

Glossary

Acid A substance which has a pH of less than 7.0. The opposite to alkali. Acidic is the word to describe something that is acid. Acid foods like lemons have a sharp or sour taste.

Alkali A substance which has a pH of more than 7.0. The opposite of an acid. Alkaline is the word to describe something that is alkali.

Breed To produce babies.

Catalytic converter A filter fitted to car exhausts to help remove pollution.

Coniferous trees Trees with cones and needle-like leaves (such as pine or fir).

Environment A plant or animal's surroundings, including air, water, soil and other plants and animals.

European Community (EC) A group of European countries which work together.

Fossil fuels Sources of energy such as coal, oil and natural gas which have been formed over thousands of years from the remains of dead animals and plants.

Minerals Substances formed naturally in rocks and the earth, such as coal and tin.

Nitrogen oxides Polluting gases formed from nitrogen in the air. They are produced when fossil fuels are burnt.

Nuclear waste Very dangerous waste material from the nuclear power industry.

Nutrients Substances that are nourishing and help a plant or animal to grow.

pH scale pH is the measure of acidity. The pH scale goes from 1 to 14 (pH1 is the most acid). There is a 10 times difference between each pH value, so pH5 is 10 times more acid than pH6.

Pollution Harmful substances in the environment.

Scandinavia The countries of Sweden, Norway and Denmark in northern Europe.

Scrubber A device attached to a factory or power station chimney to clean pollution, such as sulphur dioxide, from the fumes.

Sulphur dioxide A polluting gas formed from sulphur. It is produced when fossil fuels are burnt.

Timber Wood which is suitable for making buildings and furniture.

Picture acknowledgements
The publishers would like to thank the following for allowing their photographs to be reproduced in this book: John Baines 23 below, 28; Bruce Coleman Limited 4 (Wayne Lankinen), 5 above (John Shaw), 5 inset (Kim Taylor), 6 (John Brownlie), 8 above (Steve Kaufman), 11 (George McCarthy), 12 (Colin Molyneux), 15 (P A Hinchcliffe), 18 (John Shaw), 21 (Hans Reinhard), 23 above, 24 above (Adrian Davies), 24 below (Norman Tomalin), 25 (N Schwartz), 34 (Norman Owen), 40 (Colin Molyneux); CEGB/DIPA Photo Library 32; Greenpeace 36; ICCE 22 (Rod Redknapp), 35 above (Mark Boulton); Andre Maslennikov/IBL Sweden 16, 27; Oxford Scientific Films cover (Ronald Toms), 14 (G A MacClean), 29 (Steve Littlewood); Rex Features 37 (Boccon); Science Photo Library 30 above (Gazuit); Topham Picture Library 41; ZEFA 8 below, 10 (Hunter), 17 (Adam), 19, 20 (Mosler), 26 (M Becker), 30 below (T Braise), 31 (Bramaz), 33 (Hunter), 35 below (H Schmid), 38 (G Juckes), 42. The illustrations are by Jane Pickering 18, 22; Malcolm Walker 7; and Brian Watson 6, 9, 13, 15, 21.

Finding out more

Useful addresses

Acid Rain Foundation
1630 Blackhawk Hills
St Paul MN 55122, USA

Acid Rain Information Centre
Dept of Environment and
 Geography
Manchester Polytechnic
Chester Street
Manchester M1 5GD

Campaign for Lead-Free Air
 (CLEAR)
3 Endsleigh Street
London WC1H 0DD

Council for Environmental
 Education
University of Reading
London Road
Reading RG1 5AQ

Friends of the Earth (UK)
26-28 Underwood Street
London N1 7JQ

Friends of the Earth (Australia)
National Liaison Office
366 Smith Street
Collingwood
Victoria 3065

Friends of the Earth (Canada)
Suite 53
54 Queen Street
Ottawa KP5CS

Friends of the Earth (New
 Zealand)
Negal House
Courthouse Lane
PO Box 39/065
Auckland West

Greenpeace (UK)
30-31 Islington Green
London N1 8XE

National Society for Clean Air
136 North Street
Brighton
East Sussex BN1 1RG

Watch
22 The Green
Nettleham
Lincoln LN2 2NR

World Wide Fund for Nature
 (UK)
Panda House
Weyside Park
Godalming
Surrey GU7 1XR

Books to read

Acid Rain by Tony Hare (Franklin Watts, 1990)
Acid Rain by John Baines (Wayland, 1989)
Acid Rain by John Baines (wallchart and notes - Pictorial Charts Educational Trust, 1989)
Atmosphere by John Baines (Wayland, 1991)
The Blue Peter Green Book by Lewis Bronze, Nick Heathcote and Peter Brown (BBC Books, 1990)
The Green Detective up the Chimney by John Baines (Wayland, 1991)
The Young Green Consumer Guide by John Elkington and Julia Hailes (Gollancz, 1990)

Index

acid dust 10
acid gases 5
acid rain, damage
  to buildings
    5, 24
  to health 5
  to lakes 4, 14
  to trees 5, 14,
    20, 22, 23
acid rain, effects
  on animals 10
  on birds 18
  on fish 4, 16,
    18, 19
  on plants 10
acid snow 10, 18
acid soil 11, 15
acids 7
alkalis 7, 14, 27
Australia 6

Britain, pollution
  from 12

Canada 4, 23
  pollution in 12,
    13, 17
car exhausts 9,
  35
catalytic converter
  35, 38
chemicals 9, 10

Europe,
  pollution control
    26, 36, 38
European
  Parliament 36
experiments
  action at home 43
  how to make acid
    rain 9
  measuring
    acidity 7
  seed growth and
    acid rain 21
  soils and acid
    rain 15

fossil fuels 5, 8,
  30

Germany, pollution
  in 25, 26

hydro-electric
  power 31

Japan, pollution
  in 35, 41

lead-free petrol 35
liming 27-8, 36

minerals 15

Norway, pollution
  in 12, 13, 17

pH scale 7
pollution, control of
  26, 27-35, 36, 38
  how to reduce
    39
power stations 8,
  9, 29, 30-32

smog 25, 38
Sweden, pollution
  in 4, 13, 16, 23

USA 8, 19, 24, 30
  pollution from
    12, 13, 17
  pollution laws
    35, 38

volcanoes 8